Meditation And Astral projection

Beyond the Astral Plane

Jamie D Cook

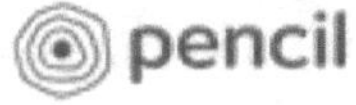

ISBN 978-93-5667-845-3
© Jamie D Cook 2023

Published in India 2023 by Pencil

A brand of
One Point Six Technologies Pvt. Ltd.
Unit no. 26, Ground Floor, Building A1,
Wadala Truck Terminal Road,
Near Post Office, Antop Hill, Mumbai - 400037
E connect@thepencilapp.com
W www.thepencilapp.com

DISCLAIMER: *The opinions expressed in this book are those of the authors and do not purport to reflect the views of the Publisher.*

Author biography

I want to thank my mom who i love and miss very much and my 7 children thank ui for your love and support and most of all God who has always been in my life and hu thru all things r possibley..............born in topeka kansas and raised in springfield missouri......Jamie D Cook is a renowned author with an impressive repertoire of books to his name. His writing style is rich and descriptive, with a lyrical quality that has captivated readers worldwide. Jamie is known for his ability to create characters that are complex and real, with flaws and strengths that make them relatable and endearing. His stories often tackle difficult and weighty issues, but Jamie handles them with a sensitivity that makes them both insightful and moving. His books incorporate elements of romance, suspense, and drama, but always with a strong emotional core that resonates with his readers. Jamie's talent has not gone unnoticed, and he has and A Know How for his writing, and he has become a sought-after speaker and panelist at literary events around the world. Despite his success, Jamie remains humble and dedicated to his craft. He is passionate about writing and is constantly exploring new ideas and themes for his next project. His fans eagerly await each new release, knowing that they will be transported into a world of beauty, complexity, and the power of the human spirit.

CONTENTS

Epigraph

"The true miracle lies not in seeking new landscapes, but in having new eyes." - Marcel Proust

Foreword

Foreword:As an avid explorer of consciousness and spirituality, I am honored to write the foreword for this transformative and insightful book on Astral Projection.The idea of Astral Projection has fascinated me for decades, and I have experienced this phenomenon numerous times throughout my spiritual journey. From my own experiences, I can attest to the profound impact that Astral Projection can have on an individual's understanding of reality and consciousness.What I appreciate most about this book is its accessible and comprehensive approach to the subject matter. The author provides readers with a step-by-step guide on how to achieve Astral Projection through meditation, while also contextualizing the practice within various spiritual and philosophical traditions. The book's combination of practical guidance and insightful commentary makes it an invaluable resource for anyone interested in exploring the depths of consciousness.Moreover, the book emphasizes the importance of safety and respect in any exploration of consciousness. It reminds us that Astral Projection is a powerful tool for self-discovery, but that it must be undertaken with care and intention. As someone who has explored the depths of consciousness through various practices, I appreciate this reminder that spiritual exploration must be approached with respect for oneself and the world around us.In summary, this book is a must-read for anyone interested in exploring the mysteries of consciousness and the universe. It offers practical guidance, personal anecdotes, and insightful commentary that will deepen your understanding and appreciation of the transformative power of Astral Projection through meditation. I highly recommend it.

Preface

Preface:Have you ever wondered what lies beyond our physical world? Do you believe that there is more to life than what we experience with our senses? If so, this book may be just what you are looking for.Astral Projection through meditation is a profound experience that offers individuals an opportunity to explore their consciousness and the universe beyond. Through meditation, individuals can learn to separate their consciousness from their physical body and travel through different planes of existence.This book offers practical guidance and advice on how to achieve Astral Projection through meditation. We cover key concepts such as intention setting, relaxation, visualization, energy work, and safety techniques, providing readers with the tools they need to explore the realms of Astral Projection safely and effectively.Furthermore, we discuss various interpretations of Astral Projection, its history, and how it relates to different spiritual and philosophical traditions. The book also includes personal accounts from individuals who have experienced Astral Projection themselves, offering readers a glimpse into the transformative and enlightening nature of this experience.Whether you are a curious beginner or an experienced practitioner of Astral Projection, this book provides a comprehensive guide that will deepen your understanding of consciousness and reality. I hope you find this book helpful and inspiring on your journey of self-discovery and exploration.

Acknowledgements

Acknowledgement:I would like to express my gratitude to everyone who contributed to the creation of this book. First and foremost, I want to thank my family and friends, who supported me throughout this journey and offered their unwavering encouragement whenever I felt lost or discouraged.I want to thank my editor and the team at the publishing house for their guidance, patience, and professionalism, turning my manuscript into a polished final product.I also want to thank the practitioners and experts in the field of Astral Projection who shared their knowledge and experiences, making this book possible. Moreover, I want to thank all the readers who choose to read and learn from this book. I hope it has provided guidance and inspiration for your own journey of Astral Projection.Finally, I would like to express my sincere gratitude to the universe and all the higher powers that may exist for guiding me on this journey of self-discovery and enlightenment.Thank you all for being part of this journey.

Introduction

21-chapter book on Meditation and Astral Projection, How to Control Astral Projection:

I. Introduction- Definition of Astral Projection and its significance- Explanation of how meditation can help co

Astral Projection?- A detailed description of Astral Projection- The barriers and prerequisites for achieving Astral Projection

III. Benefits of Astral Projection- Discuss how Astral Projection can be beneficial, including spiritual, physical, and emotional benefits

IV. Preparing for Astral Projection- Methods of physical and mental preparation for Astral Projection- The importance of relaxation and visualization

V. Different Techniques for Astral Projection- Discussion of various techniques for Astral Projection, including visualization, lucid dreaming, and meditation- Description of each technique and how they can help achieve Astral Projection

VI. The Role of Meditation in Astral Projection- Explanation of how meditation is connected to Astral Projection- The role of meditation in controlling Astral Projection

VII. Types of Meditation- Explanation of different types of meditation including mindfulness, concentration, and

ntrol Astral ProjectionII. What is

Chapter 1

As an expert in the field of Astral Projection, I can attest to the vast benefits of this practice. Astral Projection involves the separation of the astral body from the physical body, allowing one to explore different planes of existence beyond the physical plane. The significance of Astral Projection lies not only in its spiritual and mystical nature but also in its numerous benefits in terms of personal growth and well-being.While the idea of separating the astral body from the physical body may seem daunting, the process can be facilitated with the help of meditation. Meditation is a powerful tool that can help individuals control their thoughts and emotions, leading to a greater sense of awareness and inner peace. By practicing meditation regularly, individuals can train their minds to detach from their physical bodies, allowing them to enter the state necessary for Astral Projection.The combination of Astral Projection with meditation facilitates a deeper understanding of the self and the universe, leading to a greater sense of purpose and fulfillment in life. As an expert in the field, I have seen the transformative effects of Astral Projection and meditation in individuals who have successfully achieved this state. In the following chapters of this book, I will go into greater detail about Astral Projection and its benefits, techniques for preparing for and achieving Astral Projection, and how meditation can be used as a tool to control Astral Projection. With the right mindset, preparation, and practice, anyone can achieve the incredible experience of Astral Projection, leading to a more profound understanding of the universe and oneself.

Chapter2

Chapter 2 of the book provides detailed explanations of Astral Projection and the process of separating the astral body from the physical body. The chapter focuses on the barriers and prerequisites of Astral Projection, such as fear, lack of concentration, negative thoughts, and physical discomfort, and the importance of having a stable and peaceful state of mind, an open-minded approach, and a belief in oneself. The chapter provides readers with practical advice on how to overcome these barriers and meet the required prerequisites to achieve Astral Projection successfully.Chapter 2 of the book goes into great detail about what Astral Projection is and the process one goes through to achieve it. Astral Projection is described as the separating of the astral body from the physical body, allowing the individual to explore different planes of existence beyond the physical realm. During Astral Projection, individuals may experience sensations such as weightlessness, the ability to fly, and the sensation of traveling through time and space.However, achieving Astral Projection is not an easy feat and several barriers and prerequisites must be addressed. One of the primary barriers that individuals face when attempting Astral Projection is fear. The fear of detachment from the physical body can lead to an individual being unable to enter the state necessary to achieve Astral Projection.

Other barriers include lack of concentration, negative thoughts and emotions, and physical discomfort.Along with these barriers, there are also prerequisites that need to be met before attempting Astral Projection. These include a deep understanding of oneself, a stable and peaceful state of mind, an open-minded approach to exploring different planes of existence, and a belief in oneself.Therefore, it is essential to understand and overcome these barriers to accomplish Astral Projection successfully. Chapter 2 gives readers practical advice on how to address these barriers and meet the required prerequisites for Astral Projection. It teaches readers how to develop the necessary mindset and physical state to facilitate Astral Projection by eliminating these barriers.In conclusion, understanding the process and requirements for Astral Projection is the first step towards gaining the knowledge and tools to achieve it. By delving into this topic, readers will be equipped with the necessary information and tools needed to start their own Astral Projection journey.

Chapter 3

Astral Projection, also known as astral travel, refers to the ability to navigate and explore the astral plane. This can be done while the physical body is in a state of rest or sleep. Throughout the centuries, many people have believed in the transformative benefits of Astral Projection, both physically and spiritually. This chapter will explore the various benefits of Astral Projection in detail.Spiritual Benefits:For those on a spiritual journey, Astral Projection can be an effective tool. It allows one to complete a process of spiritual awakening, enabling us to become more aware of the higher levels of consciousness and the universe's interconnected nature. By accessing these higher levels of consciousness, we can develop an understanding of the interconnectedness of all beings.In Astral Projection, one can connect with their spirit guides, angels, and other entities that reside in the astral realm. Through the experience, it's possible to form deep connections with these entities, receive guidance from them, and even learn new information or skills. One can also confront negative aspects of their personality that need to be transformed and develop a much more profound connection with the spiritual realm.Physical Benefits:Astral projection can also affect the body positively. Scientifically, people who practice relaxation techniques and meditation often have lower blood pressure, a slower heart rate, and a more profound sense of relaxation. Astral projection techniques can be used to facilitate relaxation, enhance immunity, heal the body, and provide relief from various health conditions.Physical ailments such as pain and inflammation can be reduced by Astral Projection. This reduction is achieved by visualizing the problem areas in the body while in the astral realm and providing healing energy and guidance to them to aid in recovery. Many people have found it very helpful in overcoming their chronic pain conditions.Emotional Benefits:Astral Projection can be a powerful tool to manage emotions and provide mental clarity. It can help one to tap into a deeper level of consciousness where they can work on moving past negative emotions and blockages such as fear, anxiety, and anger. Astral travel teaches the mind to be more aware of our thoughts and emotional state, allowing one to develop emotional resilience and a stronger sense of self-awareness and discipline. In conclusion, the benefits of Astral Projection are endless. It's a metaphysical phenomenon that can lead to incredible spiritual growth, physical healing, and emotional balance. It's a fantastic tool to unlock human potential and connect with the universe's higher levels of consciousness.

Chapter 4

Astral Projection is an out-of-body experience where a person's consciousness travels to the astral realm while their physical body remains behind. Many people believe that Astral Projection can be a life-changing experience. However, it takes significant preparation and practice to achieve this state of consciousness. This chapter will explore the various methods of physical and mental preparation required to have a successful Astral Projection experience.Physical Preparations:The first significant step to prepare oneself for Astral Projection is to ensure there's an adequate physical space and environment conducive to relaxation. A comfortable bed is a must as Astral Projection typically occurs during sleep. Secondly, perform regular breathing exercises and light stretching to increase body relaxation.A healthy diet is essential to physical health; it's essential to avoid heavy meals before going to bed. Also, regular exercise is recommended to promote physical strength, flexibility, and endurance. Regular exercise also helps control sleep patterns, which is vital in Astral Projection.Mental Preparations:To prepare for Astral Projection, mental preparation is crucial. A state of calm and relaxation is necessary for the mind to allow for total immersion in the out-of-body experience. Here are some methods of mental preparation:1. Meditation: Meditation is one of the most effective techniques to achieve mental relaxation. It teaches the mind to focus and minimize wandering thoughts that may interfere with the Astral Projection experience. It's advised to practice meditation for at least 20 minutes daily.2. Visualization: Visualization is a powerful tool when preparing for Astral Projection. It's used to prepare the mind to envision the desired outcome of the Astral Projection experience. One can visualize the entire process and project themselves leaving their physical body and entering the astral realm.3. Affirmations: Affirmations help to imprint the desired outcome of Astral Projection in one's subconscious mind. Using affirmations such as "I can and will experience Astral Projection" helps to boost self-confidence and enhance mental clarity. Repeat the affirmations daily for optimal effectiveness.Relaxation Techniques:Deep relaxation techniques are the foundation of Astral Projection. A relaxed state is crucial to Astral Projection, as it dissociates the conscious and the subconscious mind. Here are some relaxation techniques to induce a relaxed state:1. Progressive Muscle Relaxation: This technique involves tensing and releasing groups of muscles throughout the body to achieve a deep level of relaxation.2. Breathing Techniques: Controlled and deep breathing helps to calm the mind and body. It slows down the respiration rate and helps the mind to focus.3. Guided Imagery: Guided Imagery is a form of hypnosis techniques where an expert guides the individual through mental imagery to achieve deep relaxation.In conclusion, preparing for Astral Projection requires a combination of physical and mental preparation techniques. Mental clarity, a relaxed state, and visualization of the desired outcome are key components. Practicing the above techniques can significantly improve the likelihood of having a successful Astral Projection experience.

Chapter 5

Different Techniques for Astral Projection - Discussion of various techniques for Astral Projection, including visualization, lucid dreaming, and meditation - Description of each technique and how they can help achieve Astral ProjectionAstral Projection is a state where a person's consciousness travels beyond their physical body to the astral realm. There are different techniques through which one can achieve Astral Projection. In this chapter, we will discuss some of the common techniques used to achieve Astral Projection.Visualization Technique:Visualization involves visualizing oneself leaving their body and entering the astral realm. This technique requires proper relaxation and mental concentration. To practice visualization, lie in a comfortable position in a quiet, dark room. Begin breathing deeply and slowly, and when comfortable, visualize moving your consciousness away from your physical body. Imagine yourself rising above and looking down at your physical body. Focus on the feeling of weightlessness and release of tension. Concentrate on your desire to experience astral projection. The visualization technique is suitable for beginners and anyone who wants to experience Astral Projection. However, it requires practice and concentration.Lucid Dreaming Technique:Lucid dreaming is a technique that involves being aware that you are dreaming while you are dreaming.

This technique can be used to achieve Astral Projection because you can control your dream environment and travel further into the astral realm.To practice lucid dreaming, get a comfortable night's rest. Before falling asleep, set the intention to become aware that you are dreaming. During dreaming, try to remember to question the surroundings because consciousness often follows the thought process. Slowly work on visualizing yourself floating over your physical body and into the astral realm.Lucid dreaming is a relatively easy technique for beginners. Once you have mastered the art of lucid dreaming, achieving Astral Projection becomes easier.Meditation Technique:Meditation is a powerful technique used to relax the mind and body, leading to astral travel. During meditation, the mind tends to slow down, allowing one to move beyond ordinary reality.During meditation, sit in a quiet room and focus on breathing in and out. Release all negative thoughts and visualize having a personal conversation with yourself in a safe and spiritual place. Contemplating on a repetitive sound or an image can help a practitioner experience Astral Projection.Meditation is an alternative technique for those who cannot visualize or have lucid dreams. It allows the mind to focus and reach a deeper level of consciousness, which can help initiate Astral Projection.In conclusion, there are different techniques used to achieve Astral Projection. These techniques include visualization, lucid dreaming, and meditation. Practicing and combining these techniques can increase the likelihood of a successful Astral Projection experience. While it may take some time, everyone can achieve Astral Projection with the right techniques and dedication.

Chapter 6

Meditation is a powerful tool for achieving relaxation, inner peace, and spiritual growth. There are different types of meditation, each with its unique qualities and benefits. In this chapter, we will explore the different types of meditation and how they can be useful in Astral Projection.Mindfulness Meditation:Mindfulness meditation is a type of meditation that involves paying attention to the present moment. It requires focusing on one's breath while remaining present and aware of internal and external sensations. To practice mindfulness, sit in a comfortable position with your eyes closed, and focus on your breathing. When any thoughts arise, acknowledge them, and bring your attention back to your breath. Continue for a set time or until a feeling of relaxation and calmness occurs.Mindfulness meditation can be useful in Astral Projection because it helps one to stay present and focused in the moment. This awareness can help to dissociate the consciousness from the physical body, leading to an out-of-body experience.Concentration Meditation:Concentration meditation involves focusing on a single point or object, such as a candle or a visual imagery. By focusing on a singular object, it helps to control the wandering thoughts that often interfere with meditation practice. Concentration meditation helps to improve focus, attention, and mental clarity.To practice concentration meditation, choose an object and focus your attention on it. Eventually, still the mind and concentrate on one's breath.Concentration meditation can be useful during Astral Projection because it teaches the mind to remain focused and not succumb to distractions. Astral Projection requires a high level of focus and concentration to achieve.Contemplation Meditation:Contemplation meditation involves contemplating on a specific question or idea. It requires deep thought, introspection, and self-reflection. By focusing on a particular topic, contemplation meditation helps to increase self-awareness, self-understanding, and spiritual growth.To practice contemplation meditation, choose a question or idea to contemplate. Set aside a specific time and place to focus on the topic. Meditate on the question or idea and reflect on the insights that arise.Contemplation meditation can be useful in Astral Projection because it involves introspection and self-reflection, which helps to gain a better understanding of one's subconscious desires and passions. This understanding can help lead to a more profound Astral Projection experience.Conclusion:In conclusion, there are different types of meditation, each with its unique qualities and benefits. Mindfulness meditation enhances present moment awareness and helps with relaxation, while concentration meditation improves mental focus and clarity. Contemplation meditation promotes spiritual and personal growth through self-reflection. Each meditation practice can improve one's Astral Projection experience by allowing the mind to focus, increase self-awareness, and reach a deeper state of consciousness.

Chapter 7

Meditation is an ancient practice that has become increasingly popular in today's fast-paced, modern world. It offers numerous health benefits, including reducing stress, anxiety, and depression, enhancing focus and concentration, improving sleep quality, and promoting overall mental and emotional well-being. Moreover, meditation can also be a useful tool in Astral Projection, to explore the uncharted territories of the astral realm and gain deeper insights into one's spiritual nature. In this chapter, we will explore three of the most popular types of meditation: mindfulness, concentration, and contemplation, and how each can be useful for Astral Projection.Mindfulness Meditation:Mindfulness meditation is a popular form of meditation that has gained widespread recognition in recent years. It involves being fully present in the moment, paying attention to internal and external sensations, and accepting thoughts and emotions as they arise, without judgment. Mindfulness meditation can be helpful for those new to meditation since it is relatively easy to practice and can provide a quick sense of relaxation.To practice mindfulness meditation, find a quiet place where you can sit or lie down comfortably. Close your eyes, take a few deep breaths, and then focus your attention on your breath. Observe the sensations of breathing in and out as they occur. If you find your mind wandering off, gently bring it back to your breath without judgment. When you are ready, slowly open your eyes and return to your daily activities.Mindfulness meditation can be useful in Astral Projection because it helps to quiet the mind and promote deep relaxation. This state of relaxation can make it easier to journey to the astral realm and experience an out-of-body experience.Concentration Meditation:Concentration meditation, also known as focus-based meditation or single-pointed meditation, involves focusing your attention on a single object, such as a sound, a visual image, or the breath. By concentrating your mind on one point, its wandering tendencies are reduced, and mental clarity and focus are improved.To practice concentration meditation, find a quiet and comfortable place where you can sit undisturbed. Choose an object to focus on, such as a sound or an image. Rest your attention on the object and let your thoughts pass by without becoming attached to them. If you find your mind wandering, gently bring it back to the object of your focus. Continue this practice for a set period of time.Concentration meditation can be useful in Astral Projection because it helps to develop mental focus and clarity. These qualities are vital for experiencing a successful projection, as they enable the mind to remain calm and centered during the journey to the astral realm.Contemplation Meditation:Contemplation meditation involves the deliberate contemplation of a specific idea, question or imagery. It requires engaging the intellect and emotions to derive deeper insights and understandings of oneself.To practice contemplation meditation, find a quiet and comfortable place where you can sit undisturbed. Choose a topic on which to focus and explore it thoroughly. Use all your senses to connect with the topic and allow your thoughts and emotions to unfold. Pay close attention to any insights that arise, as they may offer valuable insights into your spiritual nature.Contemplation meditation can be useful in Astral Projection because it can promote self-reflection and personal growth, which translates into a deeper understanding of one's spontaneous desires. This understanding can help one to explore the astral realm with purpose and intention, which can lead to a more profound and transformative experience.In conclusion, meditation is a powerful tool that can help to promote mental, emotional, and spiritual well-being. It can also be useful in Astral Projection, as it enables one to remain relaxed, focused, and open to new experiences. By incorporating mindfulness, concentration, and contemplation into your meditation practice, you can enhance your chances of having a successful and transformative Astral Projection experience.

Chapter 8

Guided Meditation for Astral ProjectionGuided meditation is a type of meditation that uses audio or visual cues, such as music, nature sounds, or a guided narrative, to help the practitioner enter a state of relaxation and focus. Guided meditation is particularly helpful for those who are new to meditation or struggle to remain focused on their own. In this chapter, we will explore guided meditation for Astral Projection and how it can benefit individuals who are looking to explore the inner dimensions of the astral realm.An Overview of Guided Meditation:Guided meditation can be practiced in a group setting, with a teacher or instructor guiding the participants through the meditation practice. Alternatively, guided meditations can be practiced alone, using audio or visual cues from a pre-recorded meditation. Some guided meditations focus on relaxation and stress reduction, while others are designed for enhancing mental focus, mindfulness, and personal growth.Guided Meditation for Astral Projection:Guided meditation for Astral Projection is a form of guided meditation designed to help individuals reach a deep state of relaxation and focus needed to experience an out-of-body experience. These guided meditations typically use narrative or visual prompts to create an immersive experience that leads the practitioner through the necessary steps to achieve a successful Projection.To begin, find a quiet and comfortable place where you will not be disturbed. Choose a guided meditation that is specifically designed for Astral Projection. The guided meditation will typically begin by leading you through a body scan, where you will identify areas of tension or discomfort in your body and release them through deep breathing. Next, you will be led through the steps necessary to enter the Astral realm, such as visualizing your astral body and recognizing your point of separation from your physical body.During the guided meditation, you will be asked to focus your attention on specific visualizations or sounds. Allow the guided narrative to guide you through your inner journey, and do not worry if your mind wanders, gently return your focus to the meditation when needed. The guided meditation may ask you to explore the astral realm, identify different entities, or even speak with past lives.The benefits of guided meditation for Astral Projection include helping individuals overcome the fear of the unknown and successfully inhibiting negative thoughts that can derail the projection process. Guided meditation for Astral Projection provides a way for individuals to remain focused and in the present moment, allowing for an easier transition to the astral realm.Conclusion:Guided meditation for Astral Projection can be an effective tool for those looking to explore the astral realm. Through the use of guided narrative and visualizations, individuals can enter a state of deep relaxation and focus needed to achieve an out-of-body experience. Remember that like any meditation practice, guided meditation for Astral Projection requires consistent practice and dedication to reap the full benefits. With practice, individuals can become more confident in achieving a successful Astral

Projection experience and explore the vast and mysterious inner dimensions of the astral realm.

Chapter 9

Chakra Meditation for Astral ProjectionChakra meditation is a type of meditation that focuses on the chakras, the energy centers within the body that are believed to influence physical, emotional, and spiritual health. Each of the seven chakras is associated with a specific area of the body and has its unique set of qualities. In this chapter, we will explore the role of chakras in Astral Projection and how chakra meditation can be used to enhance Astral Projection experiences.The Role of Chakras in Astral Projection:The chakra system plays an important role in Astral Projection, as it is believed that the opening and balancing of the chakras can help individuals reach a higher level of awareness and consciousness. Each of the chakras is associated with specific qualities that can affect the astral body, such as creativity, intuition, and enlightenment.By meditating on the chakras, one can better understand the energetic flows within the body, and harmonize the flow to enhance Astral Projection experiences. Activating the chakras can lead to a more profound experience as the chakras are interconnected and can enhance each other to balance the entire system.The Seven Chakras and Their Impacts on Astral Projection:There are seven chakras, each located at a different point in the body, ranging from the base of the spine to the crown of the head. Each chakra has its unique set of attributes that can impact one's Astral Projection experience:1. Root Chakra: located at the base of the spine. It is associated with grounding and stability, and aids in the awareness of the physical body. Balancing this chakra helps to achieve a stable projection and promotes a sense of safety and security in the astral realm.2. Sacral Chakra: located near the lower abdomen. It is associated with creativity and sexuality. Balancing this chakra can enhance visualization, adding to the vividness and details of astral experiences.3. Solar Plexus Chakra: located above the navel. It is associated with self-confidence and personal power. This chakra helps with maintaining the energy and focus needed during an Astral Projection and assists with overcoming the fear of the unknown.4. Heart Chakra: located in the chest. It is associated with love and compassion. A balanced heart chakra helps to cultivate a sense of peace and connection to all entities in the astral realm.5. Throat Chakra: located in the throat. It is associated with communication and self-expression. A balanced throat chakra helps one in communicating with the entities or spirits or beings encountered on the journey to the astral realm.6. Third Eye Chakra: located in the center of the forehead. It is associated with intuition and spiritual awareness. Balancing this chakra can heighten the mind's ability to imagine and enhance the ability to tune into spiritual guides and entities in the astral realm.7. Crown Chakra: located at the top of the head. It is associated with enlightenment and spiritual connection. A balanced crown chakra can help one attune to their higher self and connect to the spiritual energy surrounding them.Chakra Meditation for Astral Projection:To practice Chakra Meditation for Astral Projection, begin by finding a quiet place where you will not be disturbed. Sit or lie down in a comfortable position and focus on one of the chakras. Breathe deeply and slowly, and visualize the chakra glowing and spinning.As you focus on the chakra, see if you can sense any blocks or imbalances within your body. If you do, let that tension dissipate as you continue to breathe, and feel the energy flowing freely within the chakra. Move on to the next chakra and repeat the process until you have focused on all the chakras.Chakra meditation for Astral Projection can enhance the ability to reach a deeper level of consciousness through the harmonizing of the chakras, ultimately leading to a more profound experience.Conclusion:Meditation on the chakras can help with Astral Projection by promoting balance and harmony within the body, thus allowing the energy to flow more freely. Through consistent practice, one can become more attuned to the subtle energy flows in their body, and this can help achieve greater clarity and focus during the Astral Projection experience. As with any meditation practice, regular practice and dedication can lead to profound and transformative experiences, providing clarity, insight, and a greater connection to the astral realm.

Chapter 10

Breathing Techniques for Astral ProjectionBreathing is a vital tool in meditation practices, as it provides a gateway to the mind and body's connection. Breathing techniques can be used to help facilitate relaxation, release tension, and increase energy flow. In Astral Projection, proper breathing techniques can aid in entering a state of deep relaxation and achieving an out-of-body experience. In this chapter, we will explore different breathing techniques for Astral Projection and the importance of proper breathing during the experience.Breathing Techniques for Relaxation and Energy:Breathing techniques are often used to create a state of relaxation. They work by slowing down the breathing rate and reducing the level of stress hormones in the body. Breathing techniques can also be used to increase energy flow by increasing blood oxygenation and stimulating the nervous system's sympathetic response.Here are a few breathing techniques that can be practiced for relaxation and energy:1. Deep Belly Breathing: Inhale slowly and deeply, feeling the breath fill up the belly first and then the chest. The exhale should be slow, bringing the navel in towards the spine to help push the breath out.2. Alternate Nostril Breathing: Use the thumb and ring finger to close one nostril and inhale deeply through the other. Release that nostril and then close the other nostril and exhale through the first nostril.

Repeat the process by inhaling through the first nostril and exhaling through the second nostril.3. Kapalabhati Breathing: This is a rapid, forced exhalation through the nose that is followed by passive inhaling. This type of breathing is useful for energizing the body and clearing the mind.Importance of Proper Breathing During Astral Projection:Proper breathing techniques are essential during Astral Projection, as they help to calm the nervous system, reduce tension, and maintain focus. The correct breathing pattern can aid in the projection experience by enhancing relaxation and concentration.Shallow breathing, on the other hand, can leave individuals feeling anxious and unable to relax properly, putting a strain on the projection experience. It can also cause the body to feel sleepy or excessively charged, making it difficult to transition into an out-of-body state.Breathing during Astral Projection should be slow, deep, and rhythmic. Practicing breathing techniques before the projection can help the individual achieve a state of relaxation and ease into the projection experience.Conclusion:Breathing techniques are an essential aspect of Astral Projection, and it is crucial to practice proper breathing techniques to achieve the best results. Breathing techniques can aid in relaxation, energy flow, and focus, providing the necessary foundation for a successful projection experience. Through consistent practice of breathing techniques, coupled with intention and relaxation, individuals can achieve a deeper connection to the astral realm and explore the infinite possibilities that it holds.

Chapter 11

Exercises for Astral ProjectionVisualization and imagination are powerful tools in Astral Projection. By using visualization and imagination, individuals can create a vivid and compelling experience in the astral realm. In this chapter, we will explore the power of visualization and imagination and provide exercises to develop these skills for Astral Projection.Explanation of Visualization and Imagination:Visualization refers to the process of creating mental images or pictures in the mind's eye. It involves using the senses to imagine a specific scene or scenario. Imagination, on the other hand, involves creating something entirely new from the mind's images. It involves using creativity to generate an entirely new concept, scenario, or experience.Visualization and imagination work together to paint a vivid mental picture of the Astral Projection experience. They help to create a clear sense of what one desires to achieve during the Astral Projection.Exercises to Develop Visualization and Imagination Skills for Astral Projection:1. The Picture Exercise: Find a photo or picture that inspires you and focus on it. Study the picture in detail and then close your eyes. Try to reconstruct the image in your mind's eye, taking note of the colors, details, and textures. Practice until you can create a clear and accurate mental image.2. The Sensory Exercise: Sit in a quiet, comfortable place and close your eyes. Choose a particular object, like an apple, and try to imagine its taste, texture, and color. Let your imagination wander, and try to incorporate as many details as possible into the experience.3. The Memory Palace Exercise: Choose a familiar location, like your home or a former school, and imagine yourself walking through the space. Try to incorporate as many details as possible, such as the smells, sounds, and sights. This exercise helps to improve spatial memory and visualization skills.4. The Dream Exercise: Before going to bed, try to set an intention to have a lucid dream. When in a dream state, practice visualizing yourself consciously moving through the dream space. Use your imagination to interact with the dream objects and entities and get used to the feeling of moving through an unfamiliar space.5. The Guided Visualization Exercise: Listen to guided meditation or visualization exercises that help to cultivate the mind's visual and imaginative capacity. These exercises help to improve the ability to create detailed and vivid mental pictures.Conclusion:Visualization and imagination are powerful mental tools that can enhance the Astral Projection experience. By incorporating visualization and imagination exercises into a regular meditation practice, individuals can improve their ability to create detailed, sensory, and vivid mental pictures. Developing these skills can significantly improve the efficiency and effectiveness of Astral Projection, as it provides clarity and control over the experience. Through consistent practice, one can create an entirely new world in the mind's eye and unlock the infinite possibilities of the astral realm.

Chapter 12

Lucid Dreaming Techniques for Astral ProjectionLucid dreaming is a phenomenon in which the dreamer becomes aware that they are dreaming and can control the dream's narrative and surroundings. It is a powerful tool that can enhance the Astral Projection experience. In this chapter, we will explore the power of lucid dreaming and provide techniques to induce lucid dreaming for Astral Projection.Explanation of Lucid Dreaming:Lucid dreaming occurs when an individual is aware that they are dreaming. It allows the dreamer to control and manipulate the dream's content and surroundings. It is a natural phenomenon that occurs during the REM sleep phase, where the brain is active, and dreaming is most intense.Lucid dreaming is different from Astral Projection, as it occurs within the dream state rather than observing and interacting in an out-of-body state. However, the two experiences share many similarities, and lucid dreaming can be used as a stepping stone to Astral Projection.Techniques to Induce Lucid Dreaming for Astral Projection:1. Reality Checks: Perform routine reality checks during the day to help trigger lucid dreaming. Examples of reality checks include looking at your watch or phone to check the time, pushing your finger through the palm of your hand, or trying to read a sentence multiple times.2. Mnemonic Induction of Lucid Dreams

(MILD): This technique involves affirmations and visualization before sleep to invite the lucid dream experience. Repeat a mantra or phrase that reinforces the intention of lucid dreaming before drifting off to sleep.3. Wake-Induced Lucid Dreaming (WILD): This technique involves remaining consciously aware as the body falls asleep. Lie down in a comfortable position and focus on the hypnagogic state, the transitional period between wakefulness and sleep. This technique requires practice and may take time to master.4. Sleep Interruption Technique: Set an alarm clock to go off at a particular time during a regular sleep cycle. When the alarm goes off, stay awake for 30-60 minutes, engaging in a quiet activity like reading or meditating. This technique helps to stimulate the brain and trigger a lucid dream state.5. Dream Journaling: Keep a journal near the bed and record any dreams experienced during the night. Reflect on the dream's content, and look for recurring themes or patterns. This technique can help to improve dream recall and stimulate lucid dreaming throughout the night.Conclusion:Lucid dreaming is a powerful tool that can enhance the Astral Projection experience. By using techniques to induce lucid dreaming, individuals can improve control over their dream state and create a more meaningful and transformative dream experience. As individuals become more comfortable with lucid dreaming, they may be able to transition into the out-of-body state of Astral Projection. Through consistent practice, one can unlock the full potential of the dream and astral realm, exploring the infinite possibilities of consciousness.

Chapter 13

Mantas for Astral ProjectionMantas are powerful spiritual tools used during meditation to focus the mind and achieve deep states of consciousness. They have been used for centuries in various traditions and practices, including Astral Projection. In this chapter, we will explore what mantas are, how they work, and techniques to practice mantas for Astral Projection.Explanation of Mantas:Mantas are sacred words, phrases, or sounds that are repeated during meditation. They provide a focus point for the mind, helping individuals to enter a state of deep relaxation and concentration. By using mantas during Astral Projection, individuals can achieve a more profound connection to their higher selves and the astral realm.Mantas can be specific to a particular tradition or practice or can be chosen based on individual preferences. Examples of mantas include "Om," "So Hum," "Sat Nam," and "Aum Namah Shivaya."Techniques for Practicing Mantas for Astral Projection:1. Choose a Manta: Choose a manta that resonates with you and feels like it will assist you in achieving a connection to the astral realm. The manta should be easy to remember and have a positive vibrational frequency.2. Set the Space: Create a calm and relaxing environment in which to practice meditation and the chosen manta. This could involve lighting candles or incense, playing soft music, and sitting in a comfortable and upright position.3. Begin the Practice: Sit comfortably with your eyes closed and take a few deep breaths to relax the body and mind. Begin to repeat the chosen manta either loudly or silently in the mind. Focus solely on the manta and let it fill the mind, pushing away any other thoughts.4. Visualize: As you continue to repeat the chosen manta, allow your mind to visualize the sound waves of the manta spreading throughout your body and surrounding space. Imagine the vibrations of the manta helping to lift your consciousness out of the body and into the astral realm.5. Continue the Practice: Continue repeating the manta for as long as feels comfortable and productive. The length of time will vary based on individual practice and ability. Remember to remain relaxed and focused solely on the manta throughout the practice.Conclusion:Mantas are a powerful tool that can be used during meditation to achieve deep states of consciousness. By incorporating mantas into Astral Projection practice, individuals can achieve a more profound and meaningful connection to the astral realm. Through consistent practice, one can unlock the full potential of mantas, exploring the infinite possibilities of consciousness.

Chapter14

The Role of Attitudes and Beliefs in Astral ProjectionAstral Projection requires a specific mindset that is both open and focused. It requires positive attitudes and beliefs that support the experience and provide a foundation for exploration and growth. In this chapter, we will explore the importance of positive attitudes and beliefs in Astral Projection and provide techniques for developing them.The Importance of Positive Attitudes and Beliefs in Astral Projection:1. Belief in Possibility: Belief in the possibility of Astral Projection is essential. It provides a foundation for exploration and sets the intention for the experience. Without the belief that Astral Projection is possible, the mind may not be open enough to experience it fully.2. Fearlessness: Fearlessness is required to explore the astral realm. It allows individuals to explore and interact with the environment without fear, enabling full exploration and discovery.3. Openness: Openness is essential in Astral Projection. It allows individuals to be receptive to new experiences and knowledge, leading to growth and development.4. Trust in Self: Trust in oneself is essential for Astral Projection. It allows the individual to trust their intuition, making the experience more comfortable and enjoyable.Techniques for Developing Positive Attitudes and Beliefs:1. Visualization: Visualization is a powerful tool that can help to develop positive attitudes and beliefs. Visualize a successful and enjoyable Astral Projection experience, paying attention to details such as sights, sounds, and feelings. This technique helps to build belief in the possibility of Astral Projection and develop trust in oneself.2. Positive Affirmations: Positive affirmations are statements that help to reinforce positive attitudes and beliefs. Examples of affirmations include "I trust in my intuition," "I am fearless," and "I am open to new experiences and knowledge." Repeat these affirmations daily to develop and reinforce positive attitudes and beliefs.3. Hypnotherapy: Hypnotherapy is a form of therapy that uses guided hypnosis to access the subconscious mind. It can be used to identify and modify negative attitudes and beliefs, replacing them with positive ones that support Astral Projection.4. Journaling: Journaling is a useful technique for identifying and working through negative attitudes and beliefs. Write down any fears or doubts related to Astral Projection and explore their roots. Look for patterns and seek to replace negative beliefs with positive ones that support Astral Projection.Conclusion:Positive attitudes and beliefs are essential for Astral Projection. Belief in the possibility of Astral Projection, fearlessness, openness, and trust in oneself are all necessary components of the mindset required for a successful and enjoyable experience. Through visualization, positive affirmations, hypnotherapy, and journaling, individuals can develop and reinforce positive attitudes and beliefs that support Astral Projection. By embracing these attitudes and beliefs, individuals can unlock the full potential of Astral Projection, exploring the infinite possibilities of consciousness.

Chapter 15

Psychology of Astral ProjectionAstral Projection is a phenomenon that originates in the mind. As such, the psychology of Astral Projection plays a crucial role in understanding and experiencing this phenomenon. In this chapter, we will explore the relationship between psychology and Astral Projection, including the role of the subconscious mind in the experience.The Relationship between Psychology and Astral Projection:1. Perception: Astral Projection is a subjective experience influenced by the individual's thoughts, emotions, and beliefs. Perception plays a crucial role in how an individual experiences Astral Projection.2. Mindset: The mindset of the individual has a significant impact on the Astral Projection experience. Positive attitudes and beliefs can enhance and enrich the experience, while negative attitudes and beliefs can inhibit and limit it.3. Consciousness: The level of consciousness an individual has while Astral Projecting can vary. It is possible to achieve different levels of awareness in the astral realm, depending on the individual's mindset and level of practice.The Role of the Subconscious Mind in Astral Projection:1. Beliefs and Attitudes: The subconscious mind is responsible for beliefs and attitudes that influence the Astral Projection experience. Negative beliefs and attitudes can create fear and limit the experience, while positive ones can enhance and enrich it.2. Emotional Triggers: Emotions are often closely tied to the subconscious mind. Emotional triggers can lead to fear, anxiety, or limitations during Astral Projection. Identifying and working through emotional triggers can help to create a more positive experience.3. Visualization: Visualization is an effective tool for working with the subconscious mind. By visualizing a successful and positive Astral Projection experience, individuals can create new beliefs and attitudes that support this experience.Conclusion:The psychology of Astral Projection is essential in understanding and experiencing this phenomenon. Perception, mindset, and consciousness all play a role in the experience, and the subconscious mind can greatly influence beliefs and attitudes. By understanding these concepts, individuals can work with their subconscious mind to create a positive Astral Projection experience. With practice, individuals can achieve a deeper level of consciousness in the astral realm, unlocking the full potential of this experience.

Chapter 16

The Dos and Don'ts of Astral ProjectionAstral Projection is a fascinating and mysterious experience that requires proper knowledge and techniques to explore safely. In this chapter, we will explore precautions to take before Astral Projection and practices to avoid during Astral Projection to ensure a safe and enjoyable experience.Precautions to Take Before Astral Projection:1. Proper Mindset: Before attempting Astral Projection, it is critical to have a positive and calm mindset. Any fears or doubts should be addressed, and any distractions should be eliminated to ensure a focus on the experience.2. Safety Precautions: Ensure that the physical environment is safe before attempting Astral Projection. Remove any sharp or dangerous objects from the area and ensure that there is no possibility of someone entering the room while the individual is Astral Projecting.3. Relaxation Techniques: Astral Projection requires deep relaxation. Therefore, before attempting it, it is recommended that individuals practice relaxation techniques such as meditation, breathing exercises, or yoga.4. Practice and Patience: Like any skill, Astral Projection requires practice and patience to master. Individuals should not expect a successful experience on the first attempt and should continue to practice until they achieve their desired level of success.Practices to Avoid During Astral Projection:1. Fear and Panic: Fear and panic can be dangerous during Astral Projection. These emotions can cause the individual to abruptly return to their physical body, which can result in physical or mental harm. It is essential to remain calm and fearless during the experience.2. Interacting with Unfamiliar Spirits: It is recommended that individuals avoid contact with unfamiliar spirits during Astral Projection. These entities can be dangerous and lead to negative experiences. Additionally, it is important to remember that negative entities can sense fear, so it is essential to remain calm and confident during the experience.3. Overexertion: Overexertion can be dangerous during Astral Projection. Individuals must remember that Astral Projection is a mentally and physically exhausting experience. Therefore, it is essential to take breaks frequently and ensure that the body is adequately rested before attempting another projection.4. Alcohol and Drugs: Alcohol and drugs should be avoided before and during Astral Projection. These substances can interfere with mental and physical health and impact the quality of the experience.Conclusion:Astral Projection is an exciting and mysterious phenomenon. However, it is essential that individuals are aware of the dos and don'ts of the experience to ensure a safe and enjoyable time. Before attempting Astral Projection, individuals should have a positive mindset, take safety precautions, practice relaxation techniques, and exercise patience. During Astral Projection, individuals should avoid fear and panic, contact with unfamiliar spirits, overexertion, and avoid alcohol and drugs. By following these guidelines, individuals can have a positive and transformative Astral Projection experience.

Chapter 17

Controlling Negative Interferences During Astral ProjectionAstral Projection is a complex and highly personal experience in which the individual leaves their physical body and explores the astral plane. However, it is common for individuals to experience negative interferences during Astral Projection, which can be challenging to control. In this chapter, we will discuss negative interferences in Astral Projection and techniques for controlling them.Explanation of Negative Interferences:Negative interferences are unexpected phenomena that can occur during Astral Projection. These can range from negative entities or spirits to personal fears or tensions that the individual may hold. These interferences can cause unease and anxiety in the person, making it difficult to maintain focus and enjoy the experience. Negative interferences can be caused due to the following reasons:1. Unfamiliar surroundings: When individuals leave their physical body and enter the astral realm, they enter an unfamiliar environment. They may encounter negative entities, spirits, or negative energies, which may make them feel uneasy.2. Limiting mental state: Fear, anxiety, and limiting beliefs can also cause negative interferences. The subconscious mind can create limiting beliefs that can manifest as negative interferences during Astral Projection.3. Past Trauma: Personal experiences or negative trauma in life can affect an individual's ability to effectively Astral Project. Past trauma may cause fear or anxiety, limiting the individual's ability to control their astral experience.Techniques for Controlling Negative Interferences During Astral Projection:1. Protective Visualization: Visualization can be a powerful tool during Astral Projection. Before projecting, individuals should visualize a protective aura or shield of white light. This visualization technique will help to protect them from any negative energies or entities they may encounter.2. Positive Affirmations: Positive affirmations such as "I am safe and protected," or "I am in control," can help to build confidence and create a positive mindset during the experience.3. Eliminate Fear and Tension: Before beginning an astral projection, individuals should address any fears or tensions they may have. It is important to confront any negative beliefs and emotions to eliminate the possibility of limiting the experience.4. Proper Preparation: Proper preparation before projecting is significant. Individuals should ensure that they are adequately rested, avoid eating heavy or unhealthy foods, and clear their minds before attempting to project.Conclusion:Astral Projection is an exciting and mysterious experience that can be challenging due to the potential for negative interferences and experiences. It is essential to practice techniques to control and overcome negative interferences, such as protective visualization, positive affirmations, confronting fears and tensions, and proper preparation. With practice, individuals can maintain control over their Astral Projection experience and enjoy the fullest potential of this life-changing phenomenon.

Chapter 18

Overcoming Fear During Astral ProjectionAstral Projection is a fascinating and transformative experience that can be hindered by fear. Fear is a powerful emotion that can prevent individuals from successfully projecting and enjoying the experience. In this chapter, we will discuss how fear can prevent Astral Projection and techniques for overcoming it.Explanation of How Fear Can Prevent Astral Projection:Fear is a natural emotion that often arises when individuals attempt to Astral Project. It may be due to fears of the unknown or the potential dangers associated with leaving one's physical body. Fear can become an obstacle to Astral Projection, making it difficult for individuals to focus, relax, and let go of their physical body.Fear activates the body's "fight or flight" response, causing an increase in heart rate and rapid breathing. This response can prevent individuals from achieving a relaxed state necessary for Astral Projection. When fear sets in, individuals may attempt to fight it off, leading them to return to their body instead of projecting.Techniques for Overcoming Fear During Astral Projection:1. Practice relaxation techniques: To overcome fear during Astral Projection, it is essential to practice relaxation techniques. Breathing exercises, meditation, and yoga can help individuals to relax and achieve a calmer, more stable state of mind.2. Positive affirmations:

Affirmations can help individuals to overcome the fear associated with Astral Projection. Repeating affirmations such as "I am safe and protected" or "I am in control" can help individuals feel empowered and secure during the experience.3. Confronting fears: Confronting fears is essential to overcome them. To confront fears, individuals can create a mental image of the fear and then visualize themselves overcoming it. This technique can help individuals to feel more comfortable and less afraid during the experience.4. Create a safe environment: Creating a safe environment can help individuals to overcome their fear. To create a safe environment, individuals can create a mental image of a peaceful, comforting, and protective space where they feel safe and secure during the experience.Conclusion:Fear is a natural emotion that often arises during Astral Projection, hindering an individual's ability to achieve the experience fully. Techniques such as relaxation, positive affirmations, confronting fears, and creating a safe environment can help individuals to overcome their fears and achieve a successful Astral Projection. By practicing these techniques, individuals can overcome their fears and enjoy the full potential of this transformative experience.

Chapter 19

Astral Projection Beyond the Physical PlaneAstral Projection has tremendous potential beyond the physical plane. There are different planes of existence that individuals can explore, offering a transformative experience and deepening their understanding of the universe. In this chapter, we will discuss the different planes of existence and techniques for Astral Projection beyond the physical plane.Explanation of the Different Planes of Existence:The physical plane is the reality we experience with our physical senses. Beyond this, there exist other planes of existence that we cannot perceive with our physical senses. These planes of existence are:1. The Astral Plane: The Astral plane is the realm of emotions, thoughts, and desires. It is the realm that individuals typically access during Astral Projection. The Astral plane is where individuals encounter their spirit guides, departed loved ones, and explore their own consciousness.2. The Mental Plane: The Mental plane is the realm of thought and intellect. The Mental plane is where individuals can explore their own consciousness, connect with the collective consciousness, and develop their mental faculties.3. The Spiritual Plane: The Spiritual plane is the realm where individuals can explore their own spirituality, connect with higher beings or deities, and advance their spiritual growth.4. The Causal Plane: The

Causal plane is where individuals can explore their innate creativity and energy. This plane is where individuals can connect with their life purpose and the greater purpose of the universe.Techniques for Astral Projection Beyond the Physical Plane:1. Intention Setting: Setting a clear intention before attempting to Astral Project beyond the physical plane is essential. Individuals should have a specific goal or objective before leaving their physical body.2. Deep Relaxation: Deep relaxation is crucial when attempting to project beyond the physical plane. The deeper the relaxation, the easier it is to detach from the physical body.3. Visualization: Visualization is a powerful technique for Astral Projection beyond the physical plane. Individuals should visualize the realm they wish to visit, the beings they wish to meet, and the experiences they want to have.4. Energy Work: Energy work techniques such as chakra balancing and Kundalini yoga can help individuals to raise their energy levels and make it easier to project beyond the physical plane.Conclusion:Astral Projection beyond the physical plane offers individuals a transformative experience that can deepen their understanding of the universe. Through intention setting, deep relaxation, visualization, and energy work, individuals can achieve Astral Projection beyond the physical plane and explore the different planes of existence. As individuals expand their consciousness and explore the different planes of existence, they gain insight into their own being and the universe at large.

Chapter 20

How to Return Safely from Astral ProjectionReturning safely from Astral Projection is just as important as the experience itself. If not done properly, individuals may experience disorientation, confusion, and even physical discomfort. In this chapter, we will discuss techniques for returning safely from Astral Projection and the importance of grounding and integration.Techniques for Returning Safely from Astral Projection:1. Set a Time Limit: Before attempting to project, it is essential to have a clear time limit in mind. This will help individuals to remain focused and ensure they do not stay too long in the Astral realm. Experts recommend no more than 20-30 minutes during the initial stages of Astral Projection.2. Reconnect with the Physical Body: To return safely, individuals should focus on their physical body and visualize themselves slowly returning to it. This technique will help ensure a smooth transition from the Astral plane to the physical plane.3. Focus on Your Breath: When returning from Astral Projection, individuals may feel disoriented or anxious. Focusing on the breath can help individuals ground themselves and feel more centered before returning to their physical body.4. Use Protective Imagery: One technique for returning safely from Astral Projection is to visualize a protective cocoon around the physical body. This can help individuals feel safe and protected, allowing them to relax and gently return to their physical body.The Importance of Grounding and Integration:Grounding and integration are crucial when returning from Astral Projection. This process helps individuals to integrate the knowledge, insights, and experiences gained in the Astral realm into their physical life.1. Grounding: Grounding is essential after Astral Projection, as individuals may feel disoriented or disconnected from their physical body. To ground themselves, individuals can engage in physical activities such as walking, drinking water, or eating a small snack.2. Integration: Integration is the process of merging the knowledge, insights, and experiences gained during Astral Projection into one's physical life. This process may include journaling, meditation, or discussing the experience with a trusted friend or spiritual teacher.Conclusion:Returning safely from Astral Projection is just as important as the experience itself. By setting a time limit, focusing on the breath, using protective imagery, and engaging in grounding and integration techniques, individuals can ensure a smooth transition from the Astral realm to the physical plane. Grounding and integration help individuals to integrate the knowledge and insights gained during Astral Projection into their physical life, allowing for a transformative and meaningful experience.

Chapter 21

XXI. Conclusion

Astral Projection through meditation is a transformative experience that can deepen one's understanding of the universe and the self. In this book, we have discussed the key points for successful Astral Projection through meditation, including the importance of intention setting, relaxation, visualization, and energy work. We have also discussed the different planes of existence and techniques for returning safely from Astral Projection.

In summary, the key points for successful Astral Projection include:

- Setting a clear intention before attempting to project- Achieving a deep state of relaxation - Engaging in visualization techniques to explore different planes of existence - Using energy work techniques to raise energy levels

It is important to remember that Astral Projection is a personal experience, and what works for one individual may not work for another. It is essential to be patient and persistent, without forcing or pushing oneself too hard.

In conclusion, we recommend individuals to approach Astral Projection through meditation with an open mind, an adventurous spirit, and a willingness to explore the unknown. Through Astral Projection, individuals can gain deep insight into themselves, the world around them, and the mysteries of the universe. We hope this book has provided helpful guidance and inspiration on the journey of Astral Projection through meditation.

Appendix

Appendix - Meditation and Astral ProjectionMeditation and astral projection are two practices that are often connected with each other. While meditation is a process of training the mind to focus and clear away distracting thoughts, astral projection is the ability of the consciousness to leave the physical body and travel to other dimensions.Meditation is an effective tool for achieving a heightened state of consciousness, which can be used as a foundation for astral travel. Meditation can help to calm the mind and cultivate a sense of relaxation and detachment, which are key factors in astral projection.Astral projection is an ancient practice that has been used by many cultures around the world. The experience of astral projection can be profound, and it can provide insights and experiences that go beyond the limitations of the physical world.Before attempting astral projection, it is important to develop a strong foundation in meditation. Regular practice of meditation can help to strengthen the mind, deepen concentration, and enhance visualisation skills.During astral projection, it is important to maintain a focused and relaxed state of mind, and to have a clear intention of where you want to go and what you want to achieve. Astral projection requires a certain level of skill and it can take time to develop, so patience and dedication are important.In conclusion, meditation and astral projection are powerful practices that can provide a deeper understanding of ourselves and the universe. Through regular practice and dedication, we can develop our skills and achieve a heightened state of consciousness that can help us to explore new dimensions beyond the physical world.

Glossary

Glossary:Astral Projection - The ability to consciously leave the physical body and travel to other dimensions using the consciousness.Visualization - A technique of creating a mental image of a specific object or scenario using the imagination.Lucid Dreaming - A state of consciousness where the dreamer is aware that they are in a dream and can control its content.Mantas - A sound or phrase that is chanted or repeated during meditation, usually with a specific transformative purpose.Chakras - Centers of energy located in the body according to spiritual traditions, which are believed to impact emotional and physical health and wellbeing.Guided Meditation - A meditation performed under the guidance of a teacher or recording that provides instructions and support, usually with the purpose of achieving a specific outcome.Mindfulness - A meditation technique that focuses on paying attention to the present moment and gaining awareness and acceptance of one's thoughts and emotions.Concentration - A meditation technique that emphasizes focusing on a single point to achieve mental clarity and control over the mind.Contemplation - A meditation technique that involves deep introspection and reflection on a particular topic or question.Psychology - The study of the human mind and behavior, which can have implications for understanding experiences such as astral projection.Subconscious - The part of the mind that operates below the level of awareness, but which can influence thoughts, emotions, and behavior.Negative Interferences - Distractions or disruptions that can occur during astral projection, such as encountering dark energies or entities.Grounding - Techniques used to reconnect with the physical body and return to normal waking consciousness after astral projection.Integration - Techniques used to process and integrate the experiences and insights gained during astral projection into daily life.

Bibliography

Bibliography:1. Buhlman, William. Adventures Beyond the Body: How to Experience Out-of-Body Travel. New York: HarperOne, 1996.2. Grant, Christina. The Out-of-Body Experience: The History and Science of Astral Travel. Rochester, Vermont: Inner Traditions, 2019.3. LaBerge, Stephen. Exploring the World of Lucid Dreaming. New York: Ballantine Books, 1990.4. Lazaris. Exploring Your Inner Kingdom: Understanding and Developing the Power of Your Astral Senses. N-STAR Publishing, 1995.5. Monroe, Robert A. Journeys Out of the Body. New York: Doubleday, 1971.6. Ramacharaka, Yogi. Fourteen Lessons in Yogi Philosophy and Oriental Occultism. Kessinger Publishing, 1997.7. Roe, Joe. Astral Projection: A Beginner's Guide. CreateSpace Independent Publishing Platform, 2018.8. Tenzin Wangyal Rinpoche. The Tibetan Yogas of Dream and Sleep. Snow Lion Publications, 1998.9. Wallace, B. Alan. The Attention Revolution: Unlocking the Power of the Focused Mind. Wisdom Publications, 2006.10. Young, Michael. Astral Travel for Beginners: Transcend Time and Space with Out-of-Body Experiences. Llewellyn Publications, 2018.